Science Inquiry

Pushes and Pulls

by Joe Baron

Science Inquiry

Science in a Snap!

Push Pull Pantomime 4
Let's Move It! 5

Explore Activity

Investigate Pushes and Pulls 6

▶ **Question:** How can you change the motion of a model sailboat?

Directed Inquiry

Investigate Motion 10

▶ **Question:** How can you make a paper cup move on strings?

Math in Science 14
Estimating and Measuring Time

Directed Inquiry

Investigate Ways that Objects Can Move 18

▶ **Question:** How does the number of turns affect how fast a wind-up toy moves?

Guided Inquiry

Investigate Moving Objects 22

▶ **Question:** What happens to the motion of a ball when you blow on it?

Open Inquiry

Do Your Own Investigation 26

Think Like a Scientist

How Scientists Work 28

Inventions: Putting Things Together

Science in a Snap!

Push Pull Pantomime

Next Generation Sunshine State Standards
SC.1.P.12.1 Demonstrate and describe the various ways that objects can move, such as in a straight line, zigzag, back-and-forth, round-and-round, fast, and slow.

Pick a card. Act out what the card says. Have classmates guess if the card says to push or pull. What happens when you push? What happens when you pull?

Let's Move It!

Next Generation Sunshine State Standards
SC.1.P.12.1 Demonstrate and describe the various ways that objects can move, such as in a straight line, zigzag, back-and-forth, round-and-round, fast, and slow.

Show different ways a car can move. Have a partner tell how the car is moving. How did you get your car to move in different ways?

Explore Activity

Investigate Pushes and Pulls

Question How can you change the motion of a model sailboat?

Science Process Vocabulary

model noun

You can make and use a **model** to show how pushes and pulls work.

I can make a model of a sailboat.

observe verb

When you **observe,** you use your senses to learn about an object or event.

I observe that the foil boat is shiny.

6

Materials

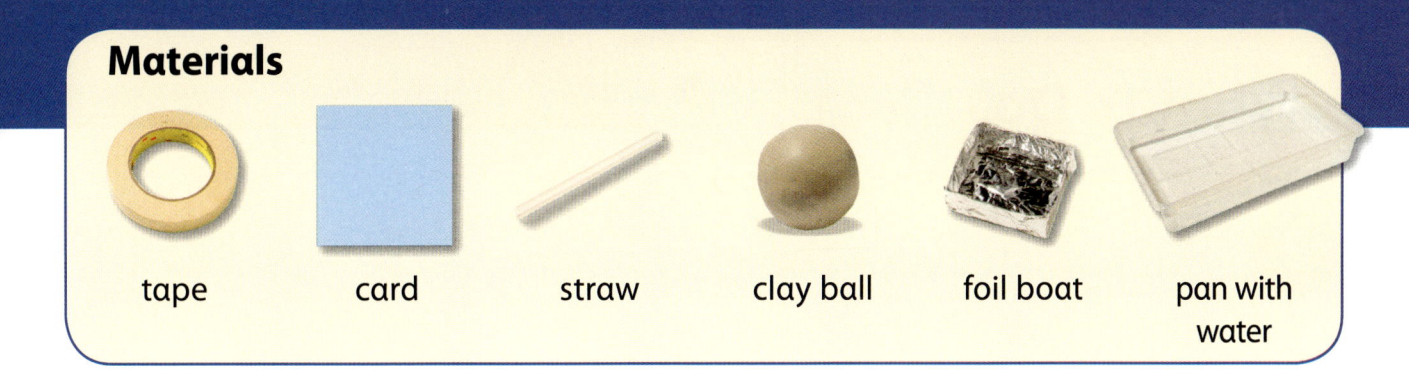

tape · card · straw · clay ball · foil boat · pan with water

What to Do

1 Make a **model** of a sailboat. First, tape the straw to the card.

2 Put the clay ball in the foil boat.

Next Generation Sunshine State Standards
SC.1.N.1.2 Using the five senses as tools, make careful observations, describe objects in terms of number, shape, texture, size, weight, color, and motion, and compare observations with others.
SC.1.P.13.1 Demonstrate that the way to change the motion of an object is by applying a push or a pull.

What to Do, continued

3 Put the straw in the clay as shown in the picture.

4 Put the boat in a pan of water. **Predict** how the boat will move if you push it or pull it. Try it. Record what you **observe** in your science notebook.

Place the boat here to start.

5 Move the boat across the pan. Use pushes and pulls to change the speed of the boat. Observe what happens.

8

Record

Write or draw a picture of what you predict and observe in your science notebook. Use tables like these.

My Science Notebook

Predict

What I Will Do	How the Boat Will Move from Starting Position
I will push the boat.	
I will pull the boat.	

Observe

What I Did	How the Boat Moved from Starting Position
I pushed the boat.	
I pulled the boat.	

Share Results

1. Tell what you did.

   *I made a **model** of _____.*

2. Tell how you moved the boat.

   I moved the boat by _____.

9

Directed Inquiry

Investigate Motion

Question How can you make a paper cup move on strings?

Science Process Vocabulary

observe verb

When you **observe,** you use your senses to learn about an object.

I observe the balloon go up.

predict verb

When you **predict,** you tell what you think will happen.

I predict that the balloon will go higher.

Materials

safety goggles 2 pieces of string cup with hole 4 rings

What to Do

1 Put on your safety goggles. Put both strings through the cup.

2 Tie a ring to the ends of each string.

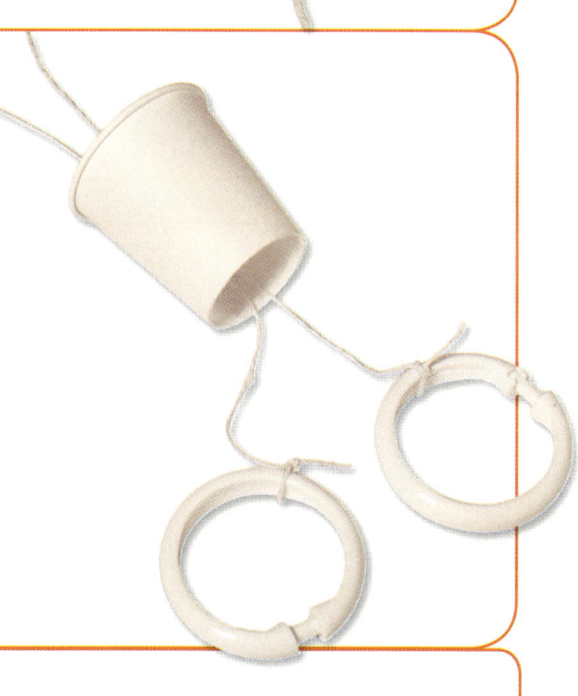

Next Generation Sunshine State Standards
SC.1.N.1.3 Keep records as appropriate, such as pictorial and written records, of investigations conducted.
SC.1.P.13.1 Demonstrate that the way to change the motion of an object is by applying a push or a pull.

What to Do, continued

3 Hold a ring in each hand. Ask a partner to hold the other two rings.

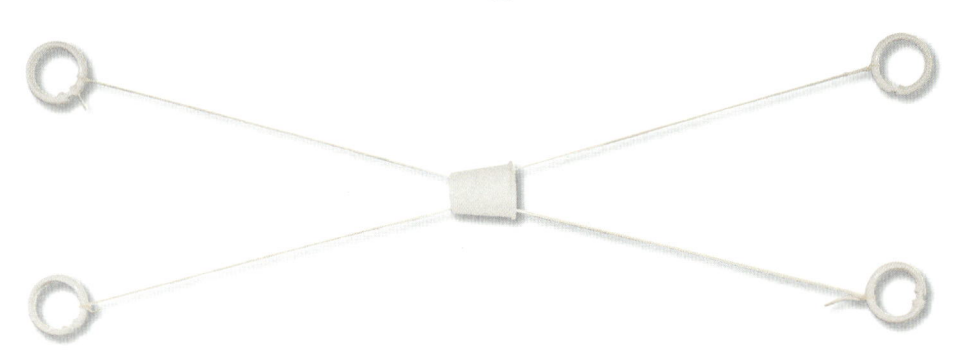

4 You can move the cup by moving the strings. **Predict** how the cup will move. Write in your science notebook.

5 Move the strings. Record your **observations.**

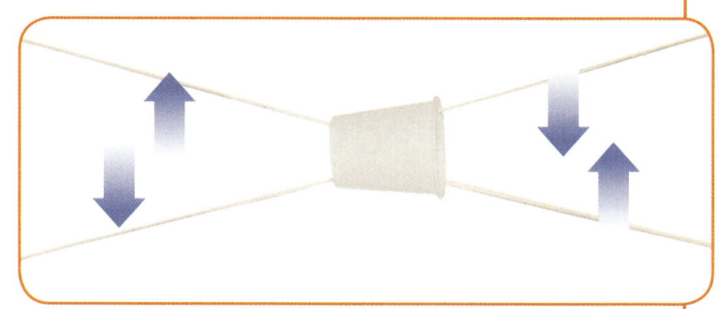

6 Discuss with others how to make the cup move fast and then slow. Try it. How can you make it stop in the middle?

12

Record

Write or draw in your science notebook. Use a table like this one.

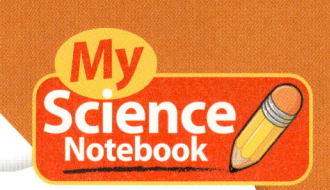

How I Moved the Ends of the String	How the Cup Moved
Apart	
Close	

Explain and Conclude

1. How did you make the cup move?
2. How did you make the cup change direction?

Think of Another Question

What else would you like to find out about how to change the direction of moving things?

13

Think Like a Scientist

Next Generation Sunshine State Standards **SC.1.N.1.3** Keep records as appropriate, such as pictorial and written records, of investigations conducted.

Math in Science

Estimating and Measuring Time

Did anyone ever ask you how long it would take to do something, such as wash your dog? How did you decide what to tell them? What you told them was an estimate.

14

Here is a game to help you estimate how long 1 minute is.

- Stand next to your desk.
- March in place when your teacher says "Start."
- Stop marching when your teacher says "Stop."
- Think about how much time you marched.

You marched for 1 minute! Now you know how long a minute is. You can make better estimates!

Think Like a Scientist

continued

You can use a stopwatch to measure time. A stopwatch shows the number of hours, minutes, and seconds.

On the stopwatch, 0:01'23" means 1 minute and 23 seconds.

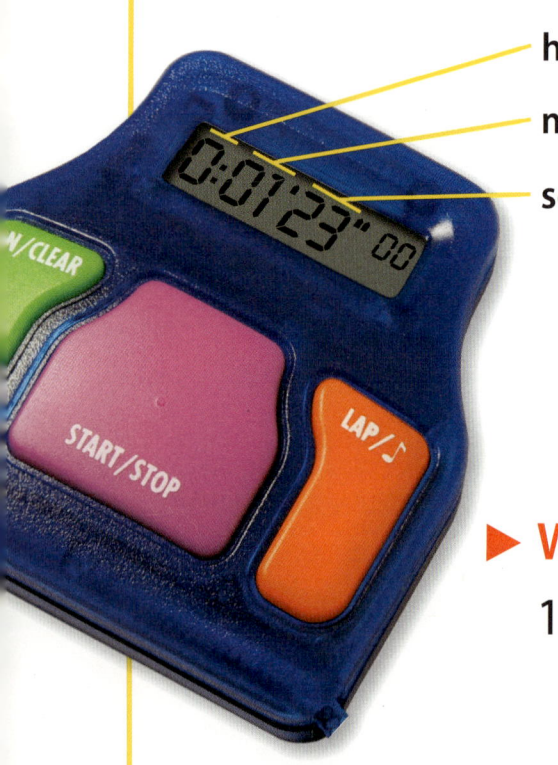

- hours
- minutes
- seconds

▶ **What Did You Find Out?**
1. Your stopwatch shows 0:00'55". Is that time more or less than 1 minute? How much time is it?
2. How might knowing how long a minute is help you make better estimates? Explain.
3. When might you use this information?

Estimate and Measure Time

1. Estimate and measure how long it will take a ball to roll 2 meters. Work with a partner.

   - Make a time estimate.
   - Have your partner start the stopwatch and say "Start." Slowly push the ball at the same time.
   - Stop the stopwatch when the ball gets to the end of the path.

2. Estimate and measure 2 more times.
3. How close was your estimate to the actual time?

Directed Inquiry

Next Generation Sunshine State Standards
SC.1.N.1.3 Keep records as appropriate, such as pictorial and written records, of investigations conducted.
SC.1.N.1.4 Ask "how do you know?" in appropriate situations.
SC.1.P.12.1 Demonstrate and describe the various ways that objects can move, such as in a straight line, zigzag, back-and-forth, round-and-round, fast, and slow.

Investigate Ways that Objects Can Move

Question How does the number of turns affect how fast a wind-up toy moves?

Science Process Vocabulary

predict verb

When you **predict**, you say what you think will happen.

I predict that the toy will move.

data noun

You collect and record **data** when you gather information in an investigation.

I can use a meterstick to collect data.

18

Materials

meterstick · tape · wind-up toy · stopwatch

What to Do

1 Put a piece of tape on the floor. Use a meterstick to **measure** 30 centimeters from the tape. Place another piece of tape there.

2 Predict. How does the number of turns change how fast a wind-up toy moves? Write in your science notebook.

19

What to Do, continued

3 Turn the knob 5 times. Place the toy on the **Start** tape.

4 Use a stopwatch. Measure how long it takes the toy to get to the **End** tape. Write the **data** in your science notebook.

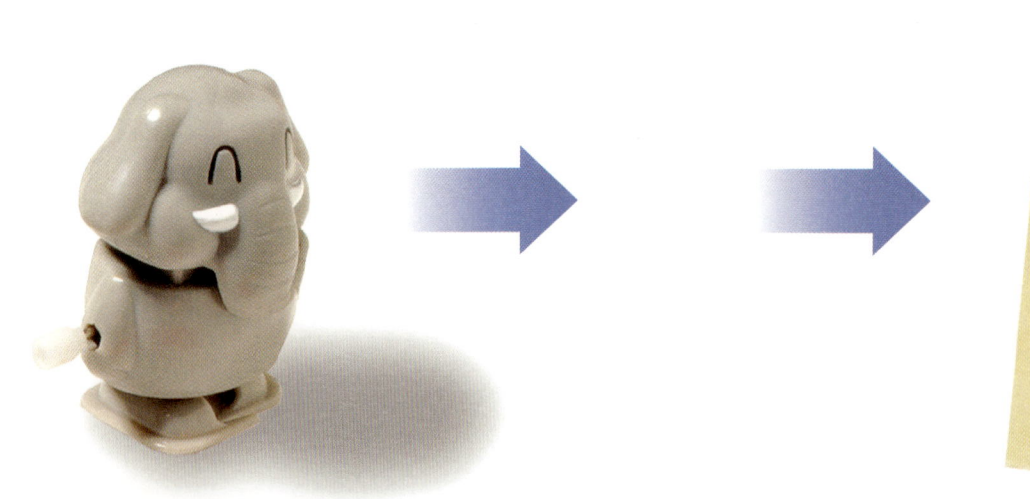

5 Repeat steps 3 and 4. This time, turn the knob 7 times.

20

Record

Write in your science notebook. Use a table like this one.

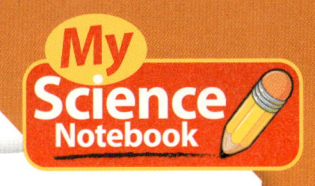

Number of Turns	Time It Took to Go 30 cm (seconds)
5	
7	

Explain and Conclude

1. Did the results support your **prediction?** Explain.
2. How did the number of knob turns change how fast the toy moved?

Think of Another Question

What else would you like to find out about how you can make a wind-up toy move? What could you do to answer this new question?

Guided Inquiry

Investigate Moving Objects

Question What happens to the motion of a ball when you blow on it?

Science Process Vocabulary

infer verb

When you want to explain something that you have seen, you can **infer**.

I infer that wind moves the flag.

share verb

You can **share** results by writing or talking about what happens.

22

Materials

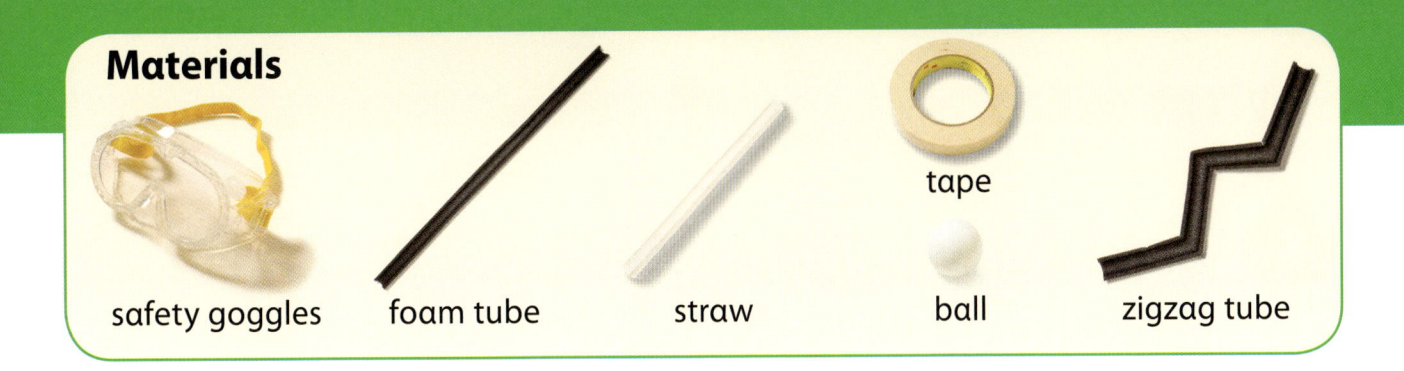

safety goggles | foam tube | straw | tape | ball | zigzag tube

What to Do

1 Tape the ends of the straight tube to the floor.

2 Put on your safety goggles. Put a ball at the end of the tube. Use the straw to blow the ball along the tube.

Next Generation Sunshine State Standards
SC.1.N.1.4 Ask "how do you know?" in appropriate situations.
SC.1.P.12.1 Demonstrate and describe the various ways that objects can move, such as in a straight line, zigzag, back-and-forth, round-and-round, fast, and slow.
SC.1.P.13.1 Demonstrate that the way to change the motion of an object is by applying a push or a pull.

What to Do, continued

3 Make the ball go back-and-forth. Blow the ball to one end of the tube. Then blow the ball back to the other end.

4 Form the tube into a circle. Tape the ends together. Blow the ball round-and-round inside the circle.

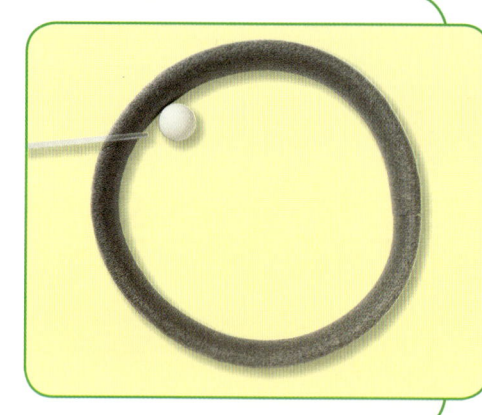

5 Use the zigzag tube. Blow the ball along the path.

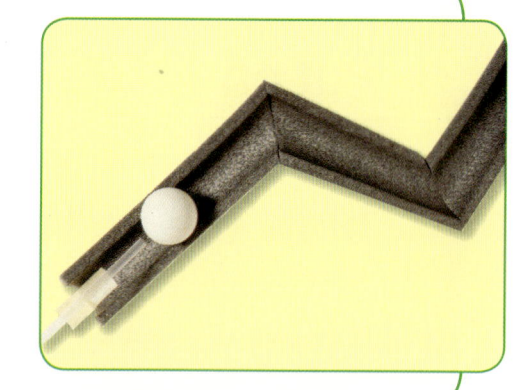

6 Tape together pieces of foam tube to make a new path. Blow the ball along your path.

7 **Infer** what changed the motion of the ball in each step. Write in your science notebook.

Record

Draw pictures to show the motion of the ball. Infer what changed the ball's motion.

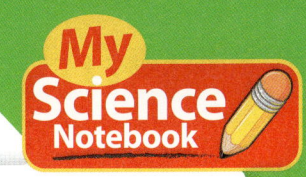

Type of Motion	What I Infer Changed the Motion

Explain and Conclude

1. **Compare** the motion of the ball in steps 2 and 5.
2. **Share** results. Tell how the motion of the ball changed in step 6.
3. What kind of force changed the motion of the ball in each step?

Think of Another Question

What else would you like to find out about how objects move in different ways?

25

Open Inquiry

Do Your Own Investigation

Question Choose a question, or make up one of your own to do your investigation.
- How can you change the direction of a paper airplane?
- How can you change how a bubble moves through the air?

Science Process Vocabulary

fair test

In **fair test** you change only one thing.

I will only change where I put the paper clip on the paper airplane.

Open Inquiry Checklist

Here is a checklist you can use when you investigate.

Next Generation Sunshine State Standards
SC.1.N.1.1 Raise questions about the natural world, investigate them in teams through free exploration, and generate appropriate explanations based on those explorations.

- ☐ Choose a **question** or make up one of your own.

- ☐ Gather the materials you will use.

- ☐ Tell what you **predict.**

- ☐ Plan a **fair test.**

- ☐ Make a **plan** for your investigation.

- ☐ Carry out your plan.

- ☐ Collect and record **data.** Look for **patterns** in your data.

- ☐ Explain and **share** your results.

- ☐ Tell what you **conclude.**

- ☐ Think of another question.

27

Think Like a Scientist

Next Generation Sunshine State Standards **SC.1.N.1.3** Keep records as appropriate, such as pictorial and written records, of investigations conducted. **SC.1.P.12.1** Demonstrate and describe the various ways that objects can move, such as in a straight line, zigzag, back-and-forth, round-and-round, fast, and slow.

How Scientists Work

Inventions: Putting Things Together

Look at the parts in the picture. How would you put the parts together to invent a racer? Think about what the parts can do together that they cannot do alone.

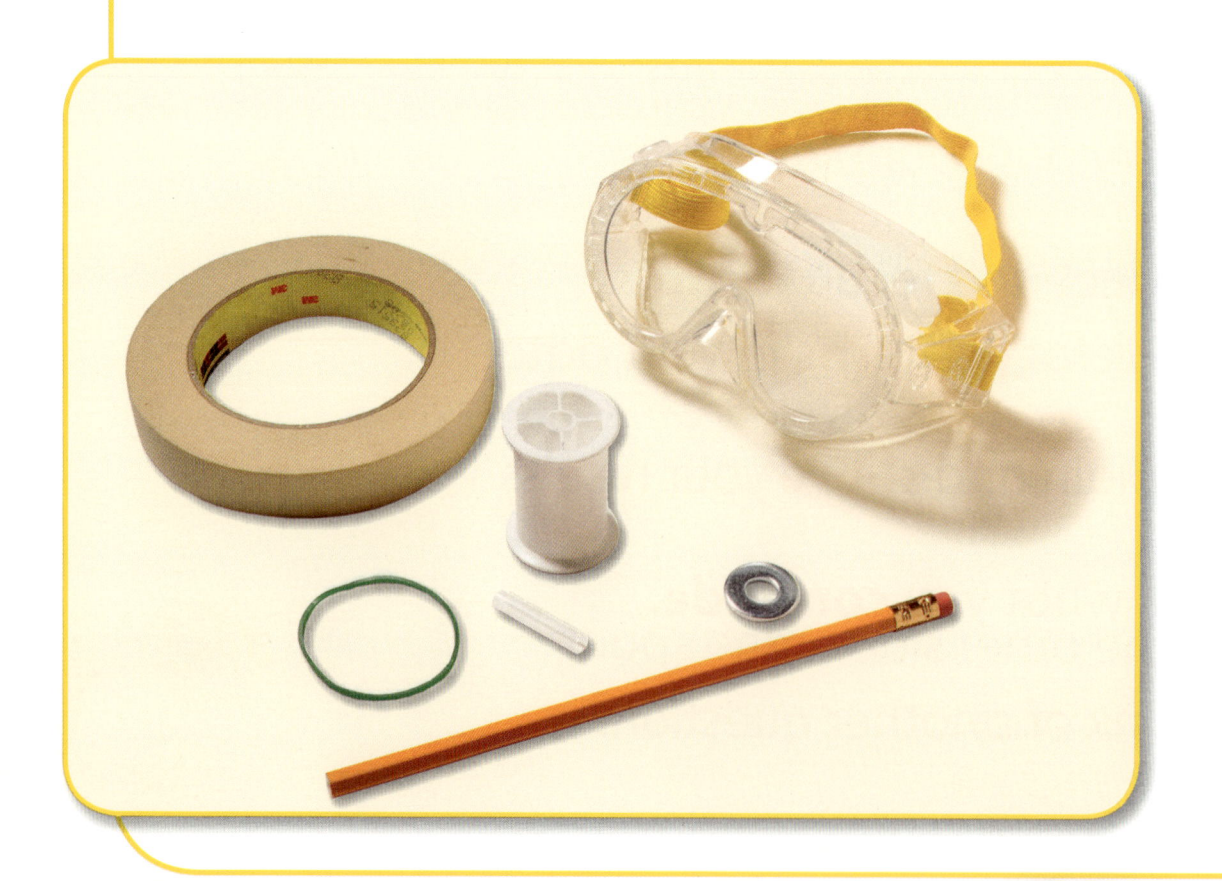

Here is one way you might use the parts in the picture to make a racer.

1. Put on your safety goggles. Then slip the rubber band in the hole of the spool.

2. Put the piece of straw in one loop of the rubber band.

3. Tape the straw to the spool.

4. Slip the washer over the loop on the other side of the spool. Put the pencil in the loop.

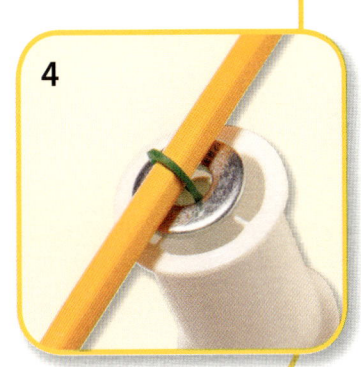

Think Like a Scientist

continued

You have made a racer! Now you need to test it. Twist the pencil 20 times. Place the racer on the floor. Watch it move across the floor! Test it 3 more times to see how it moves.

▶ **What Did You Find Out?**
1. What could the parts do together that they could not do alone?
2. What made the racer move?

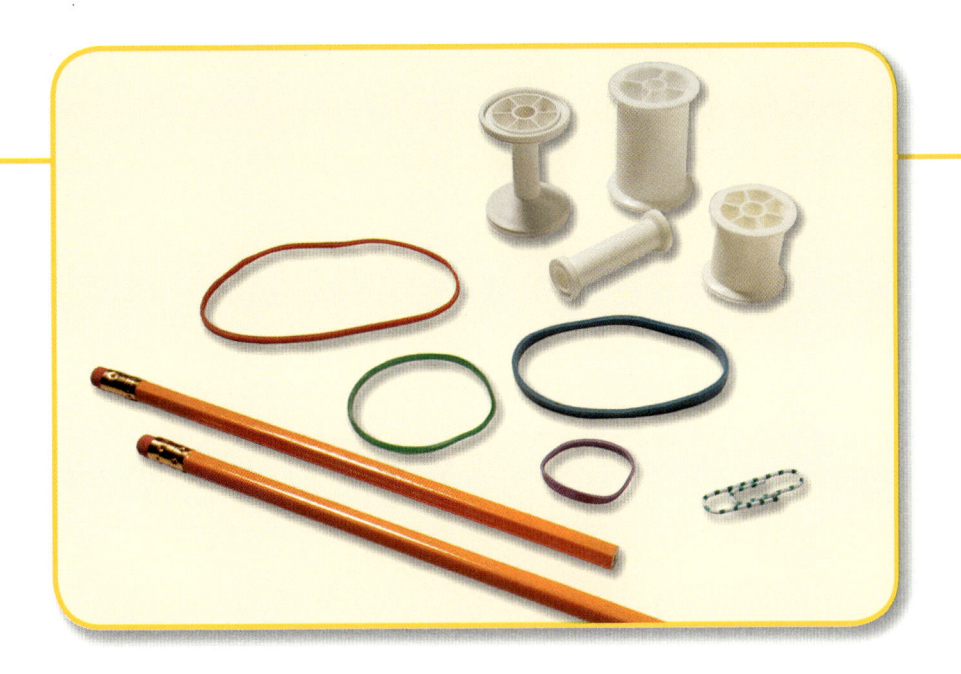

✋ Put Things Together

1. Look at the parts above. What parts can you use to make your racer better? Can you make it go faster? Can you make it go farther?

   - What parts will you use to change your racer?

   - How will you put the parts together? List the steps.

2. Try your new racer again and again. How does the racer move now?

31

Featured Photos

Cover: line of men pulling in boat from sea, Bali, Indonesia

Title page: pulling sled up a snowy hillside

inside back cover: tractor

ACKNOWLEDGMENTS
Grateful acknowledgment is given to the authors, artists, photographers, museums, publishers, and agents for permission to reprint copyrighted material. Every effort has been made to secure the appropriate permission. If any omissions have been made or if corrections are required, please contact the Publisher.

PHOTOGRAPHIC CREDITS
set-up photography: Andrew Northrup; stock photography; Cover (bg) ImageState/Pictor International/Alamy Images; Title (bg) Hans Huber/Westend61/Jupiterimages; 5 (t) HomeStudio/Shutterstock; 6 (t) Giovanni Rinaldi/iStockphoto; 6 (cr) Ricardo De Mattos/iStockphoto; 10 (t) Nikada/iStockphoto; 10 (b) Fancy/Veer/Corbis; 13 (b) suemack/iStockphoto; 14 (t) Michael Keller/Corbis; 14 (b) Judith Collins/Alamy Images; 18 (t) RFStock/Alamy Images; 18 (c) Diane Diederich/iStockphoto; 21 (l) PhotoSpin, Inc/Alamy Images; 21 (r) maxstockphoto/Shutterstock; 22 (t) Suzanne Tucker/Shutterstock; 22 (c) PhotoDisc/Getty Images; 22 (b) Ariel Skelley/Blend Images/Jupiterimages; 25 (b) Lester Lefkowitz/Corbis; 26 (t) Phil Degginger/Alamy Images; 26 (b) PhotoDisc/Getty Images; 29 (b) paul ridsdale/Alamy Images; 30 (t) Ace Stock Limited/Alamy Images; 30 (b) ELC/Alamy Images; 31 (b) Susanna Price/Dorling Kindersley/Getty Images; Inside Back Cover Nigel Cattlin/Alamy Images.

Neither the Publisher nor the authors shall be liable for any damage that may be caused or sustained or result from conducting any of the activities in this publication without specifically following instructions, undertaking the activities without proper supervision, or failing to comply with the cautions contained herein.

PROGRAM AUTHORS
Judith Sweeney Lederman, Ph.D., Director of Teacher Education and Associate Professor of Science Education, Department of Mathematics and Science Education, Illinois Institute of Technology, Chicago, Illinois; Randy Bell, Ph.D., Associate Professor of Science Education, University of Virginia, Charlottesville, Virginia; Malcolm B. Butler, Ph.D., Associate Professor of Science Education, University of South Florida, St. Petersburg, Florida; Kathy Cabe Trundle, Ph.D., Associate Professor of Early Childhood Science Education, The Ohio State University, Columbus, Ohio; Nell K. Duke, Ed.D., Co-Director of the Literacy Achievement Research Center and Professor of Teacher Education and Educational Psychology, Michigan State University, East Lansing, Michigan; David W. Moore, Ph.D., Professor of Education, College of Teacher Education and Leadership, Arizona State University, Tempe, Arizona

THE NATIONAL GEOGRAPHIC SOCIETY
John M. Fahey, Jr., President & Chief Executive Officer
Gilbert M. Grosvenor, Chairman of the Board

Copyright © 2011 The Hampton-Brown Company, Inc., a wholly owned subsidiary of the National Geographic Society, publishing under the imprints National Geographic School Publishing and Hampton-Brown.

All rights reserved. No part of this book may be reproduced or transmitted in any form or by any means, electronic or mechanical, including photocopying, recording, or by an information storage and retrieval system, without permission in writing from the Publisher.

National Geographic and the Yellow Border are registered trademarks of the National Geographic Society.

National Geographic School Publishing
Hampton-Brown
www.NGSP.com

Printed in the U.S.A.
RR Donnelley, Johnson City, TN

ISBN: 978-0-7362-7646-7

11 12 13 14 15 16 17

10 9 8 7 6 5 4